CRITICAL THINKING & LOGICAL REASONING WORKBOOK-7

7

GIFT OF LOGIC™ SERIES

An Essential Resource for Everyone

Boost Your Thinking Skills

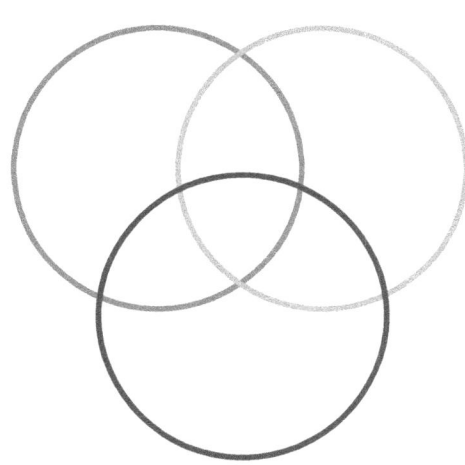

Verbal Reasoning

Analytical Reasoning

Pictorial Reasoning

THIRD EDITION

| FOR GRADES 6-12 | STUDENTS, TEACHERS, AND PARENTS |

Ranga Raghuram

GIFT OF LOGIC™

Gift Of Logic, Inc

http://www.giftoflogic.com
sales@giftoflogic.com

Critical Thinking and Logical Reasoning Workbook-7
ISBN-13: 978-1494832551
ISBN-10: 1494832550

Third Edition
1-2014

Copyright © 2009 Gift Of Logic, Inc. All rights reserved. No part of this publication may be reproduced, stored in a retrieval system, transmitted in any form or by any means, electronic, mechanical, photocopying, recording or otherwise, without the written permission of the publisher.

License: This book is licensed for use by one person only. Use of this book in a group setting (classroom, workshop, etc) without the written permission of the publisher is prohibited. Unauthorized duplication is strictly prohibited by law. Contact the publisher at sales@giftoflogic.com for classroom/school/group licensing.

GIFT OF LOGIC™
CRITICAL THINKING & LOGICAL REASONING CURRICULUM
12 WORKBOOKS TO BOOST YOUR THINKING SKILLS

For Kindergarten, Grade 1, and Grade 2

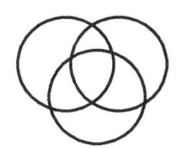

Workbook# 0

Verbal Reasoning	Finding the truth, Inferencing, Analogies, Synonyms and Antonyms, Agree/Disagree
Analytic Reasoning	Memory drill, Decision making, Positioning, Sudoku
Pictorial Reasoning	Connect the dots, Mazes, Picture Sequence, Spot the difference, etc

Workbook# 1

Verbal Reasoning	Finding the truth, Inferencing, Analogies, Synonyms and Antonyms, Agree/Disagree
Analytic Reasoning	Sorting, Positioning, Picking, Assorted problems, Numeric and Alphabetic Sudoku
Pictorial Reasoning	Picture Sequence, Spot the difference, Odd picture

Workbook# 2

Verbal Reasoning	Finding the truth, Classification, Direct and Inverse relationship, Inferencing, Analogies, Agree/Disagree
Analytic Reasoning	Sequencing, Scheduling, Strategy, Picking, etc
Pictorial Reasoning	Picture Analogy, Odd picture, Pattern matching, etc

For Grade 3, Grade 4, and Grade 5

Workbook# 3

Verbal Reasoning	Not, And, Or, If .. then, Conditional inferencing, Unconditional inferencing, Symbolic Logic
Analytic Reasoning	Lists, Sequencing, Grouping, Venn Diagrams, Graph logic, Number logic, Letter logic, Sudoku
Pictorial Reasoning	Picture sequence, Picture analogy, Odd picture, Picture difference, Pattern matching

Workbook# 4

Verbal Reasoning	Contradiction, Converse, Inverse, Contrapositive, Conditional inferencing, Symbolic Logic
Analytic Reasoning	Scheduling, Looping, FIFO, LIFO, Correlation, Venn Diagram, Graph logic, Number logic, Sudoku, etc
Pictorial Reasoning	Picture sequence, Picture analogy, Odd picture, Picture difference, Pattern matching

Workbook# 5

Verbal Reasoning	Biconditional, Categorical inferencing, Cause and Effect, Symbolic Logic, Agree/Disagree, Word and Sentence analogy
Analytic Reasoning	Correlation, Grouping, Venn Diagrams, Graph logic, Number logic, Letter logic, Sudoku, etc
Pictorial Reasoning	Picture sequence, Picture analogy, Odd picture, Picture difference, Pattern matching

********* Essential resource for everyone *********
*http://www.giftoflogic.com *sales@giftoflogic.com

GIFT OF LOGIC™
CRITICAL THINKING & LOGICAL REASONING CURRICULUM
12 WORKBOOKS TO BOOST YOUR THINKING SKILLS

For Grades 6-12, College/University Students, Adults

Primer

Prereq

Verbal Reasoning	Logical Operators, Conditional, Categorical and Causal reasoning, Validity, Fallacies, Symbolic Logic
Analytic Reasoning	Positioning, Grouping, Sudoku
Pictorial Reasoning	Pattern perception, Figure formation, Paper folding and cutting, Figure matrix, Rule detection

Workbook# 6

Verbal Reasoning	Arguments-Main point, Must be true, Cannot be true
Analytic Reasoning	Positioning, Grouping, Sudoku
Pictorial Reasoning	Pattern perception, Figure formation, Paper folding and cutting, Figure matrix, Rule detection

Workbook# 7

Verbal Reasoning	Arguments-Strengthening, Weakening
Analytic Reasoning	Positioning, Grouping, Sudoku
Pictorial Reasoning	Pattern perception, Figure formation, Paper folding and cutting, Figure matrix, Rule detection

Workbook# 8

Verbal Reasoning	Arguments - Controversy, Paradox
Analytic Reasoning	Positioning, Grouping, Sudoku
Pictorial Reasoning	Pattern perception, Figure formation, Paper folding and cutting, Figure matrix, Rule detection

Workbook# 9

Verbal Reasoning	Arguments- Assumptions, Reasoning strategy
Analytic Reasoning	Positioning, Grouping, Sudoku
Pictorial Reasoning	Pattern perception, Figure formation, Paper folding and cutting, Figure matrix, Rule detection

Workbook# 10

Verbal Reasoning	Arguments-Flawed reasoning, Analogous reasoning
Analytic Reasoning	Positioning, Grouping, Sudoku
Pictorial Reasoning	Pattern perception, Figure formation, Paper folding and cutting, Figure matrix, Rule detection

********* Essential resource for everyone *********
Get the GIFT OF LOGIC™ today !
*http://www.giftoflogic.com   *sales@giftoflogic.com

© Gift Of Logic, Inc * Copying prohibited

Dear Reader:

Your decision to purchase this book is commendable. You now have in your hands, a comprehensive, easy-to-read book in Critical thinking and Logical reasoning that will introduce you to three different areas of thinking and reasoning - Verbal, Analytical and Pictorial. Solving problems in Verbal Reasoning is important to develop a critical mind. Solving problems in Analytic Reasoning is important to develop a flexible and resourceful mind. Solving problems in Pictorial Reasoning is important to develop a visually alert mind.

This book is presented in a workbook format to help you progress quickly. Parents and teachers are urged to complete the exercises ahead of the student and assist them whenever necessary with the help of detailed answers provided at the end of the book. This book can be used as a supplementary resource in the regular class room or it can be used during winter and summer vacations. College/University students, working professionals and retired individuals will also find the Gift Of Logic(tm) Series very useful in enhancing their problem solving abilities, confidence and general intellect.

Critical thinking and Logical reasoning must be practiced consistently to develop strong cognitive skills. After completing the exercises in this book, continue to read the other books in this series to get familiar with different types of Logical reasoning problems.

This workbook is one in a series of twelve workbooks. Please refer to the brochure before this page for a brief description of each workbook. Visit the website http://www.giftoflogic.com for more information.

                                                           Happy thinking and reasoning!

TABLE OF CONTENTS

Verbal Reasoning

Strengthen... 8
Weaken.. 16

Analytical Reasoning

Sudoku... 26
Positioning... 31
Grouping.. 43

Pictorial Reasoning

Patter perception.. 52
Figure formation... 54
Paper folding and cutting... 56
Figure matrix.. 57
Rule detection.. 59

Answers

Verbal... 62
Analytic.. 91
Pictorial.. 117

Certificate of Completion

Name _____   Date _____

VERBAL REASONING

STRENGTHEN

In this section on "Strengthen", you will develop the ability to strengthen an argument.

A complete argument is given (premises and conclusion) and your task is to pick a premise that strengthens the argument.

After the argument is presented, you will be posed questions such as the following:
 *Which one of the following statements most strengthens the argument?
 *Which one of the following most strongly supports the argument?

The correct answer is the one that strengthens the argument the most. That is, if two answer choices both strengthen the argument, then the one that strengthens the most is the correct answer. For example, if the conclusion of an argument is "Therefore, it is likely to rain today", then the statement "There is a thirty percent chance of rain today" strengthens the argument more strongly than the statement "There is a ten percent chance of rain today".

Incorrect answers weaken the argument instead of strengthening it. For example, if the conclusion of an argument is "Therefore, it is likely to rain today", then the statement "There are no dark clouds in the sky", weakens the argument instead of strengthening it.

1	STRENGTHEN	Weather

Exactly one hundred days ago, the temperature was very cold. Therefore, we can conclude that it rained that day.

Which one of the following, if true, most strengthens the argument?

A) On that day, there were many dark clouds in the sky.
B) On that day, there was bright sunshine.

2	STRENGTHEN	Garden

Tomatoes in several plants in Joe's garden were found in a half-bitten state. Joe knew that there were rabbits in the garden. So, Joe was sure that these rabbits were responsible for the half bitten tomatoes.

Which one of the following, if true, most strengthens the argument?

A) Joe's sister Amanda saw rabbits chewing the tomatoes.
B) Joe's garden is protected by an alarm to scare the rabbits.

Name _____   Date _____

| 3 | STRENGTHEN | Family |

Laura is the person responsible for arranging a family reunion in a hotel. She sent out invitations to all of her relatives and expected an overwhelming response. But, only ten members expressed willingness to come to the reunion. Therefore, Laura decided to cancel the reunion.

Which one of the following, if true, most strengthens the argument?

A) The hotel requires at least five people to attend a reunion.
B) Holding a reunion will make sense only if all the family members attend.

| 4 | STRENGTHEN | Accidents |

Seventy percent of the road accidents happen because of drivers speeding over the limits. The remaining thirty percent of the accidents occur because of other reasons. Therefore, to prevent a majority of accidents, we must increase the fines on drivers speeding over the limits.

Which one of the following, if true, strongly supports the argument?

A) Fines do not have an influence on driver behavior.
B) Drivers do not like to pay fines.

Name _____   Date _____

| 5 | STRENGTHEN | Causal |

Farmer Joe noted nothing unusual in his small farm before retiring for the day. The next morning, he was very surprised to see a one foot wide trench running across his farm. Since there was a trembling movement last night, he concluded that an earthquake caused the crack in his farmland.

Which one of the following, if true, most strengthens the argument?

A) Earth in farmland can crack suddenly due to insufficient irrigation.
B) Seismographs in the area recorded an earthquake last night.

| 6 | STRENGTHEN | Farmers |

Apple farmers bring their produce to the market to sell them to customers. Last year, they met the high demand for apples and reaped a handsome profit from the sale. This year too, in spite of reports of worms inside the apples, the high demand was met. Therefore, the apples brought to the market this year are safe to eat.

Which one of the following, if true, most strengthens the argument?

A) Food inspectors tested the apples and found worms inside.
B) Food inspectors tested the apples and did not find worms inside.

7	STRENGTHEN	Actors

If an actor is wealthy, he will donate to charities. Therefore, if an actor is wealthy, he will be popular.

Which one of the following, if true, most strengthens the argument?

A) If an actor is popular, he will donate.
B) If an actor donates, he will be popular.

8	STRENGTHEN	Sports

Soccer is a game that requires good athletic abilities. One has to be constantly on the move in the playground during the game. This can tire out the players soon-both boys and girls. A sports-drink manufacturing company found out that girls tire sooner than boys when they consume Drink-A during the game, but boys tire sooner than girls when they consume Drink-B during the game. So, the company decided to sell Drink-A for boys and Drink-B for girls.

Which one of the following, if true, most strengthens the argument?

A) Drink-A and Drink-B have the same ingredients.
B) Drink-A is specially formulated for males and Drink-B is specially formulated for females.
C) Male dogs tire out sooner than female dogs when they consume Drink-B and female dogs tire out sooner than male dogs when they consume Drink-A.

Name _____   Date _____

| 9 | STRENGTHEN | Sports |

The coach of a defeated team tried to figure out why his team lost a crucial game. He could think of only two possibilities for his team's dismal performance-either a lack of adequate practice or a lack of team spirit. It was obvious that there was lack of coordination among the team members during the game. This led him to believe that he should introduce more team building activities.

Which one of the following, if true, strongly supports the argument?

A) The winning team showed better team spirit during the game.
B) The defeated team members argued about their roles during the game.

| 10 | STRENGTHEN | Car |

Brendan had the option of buying the blue car or the red car. The blue car was cheaper than the red car. So, Brendan bought the blue car. But, he regretted his decision soon as the blue car broke down several times leading to very expensive repairs. So, he decided to sell his blue car and buy the red car.

Which one of the following, if true, strongly supports the argument?

A) The red car is known to be less problematic than the blue car.
B) The blue car consumes less fuel than the red car.

Name _____ Date _____

| 11 | STRENGTHEN | Medical |

Two types of shots were offered by a medical clinic, Type-A and Type-B. Type-A shots were supposed to prevent bronchitis. Type-B shots were supposed to prevent pneumonia. Linus was given both the shots, but that did not stop him from contracting both bronchitis and pneumonia. So, he came to the conclusion that the shots were useless.

Which one of the following, if true, strongly supports the argument?

A) The shots helped several people resist bronchitis and pneumonia.
B) Several people who took both the shots reported no resistance to bronchitis and pneumonia.

| 12 | STRENGTHEN | Psychology |

Psychologist: When people make small mistakes, they report it immediately. But, when they make big mistakes, they neither report it nor do they tell the truth when questioned. This behavioral pattern leads us to the conclusion that people are not honest at all times.

Which one of the following, if true, strongly supports the argument?

A) People are not afraid of reporting small mistakes.
B) People are ashamed to admit making big mistakes.

Name _____   Date _____

| 13 | STRENGTHEN | University |

The University that Michael wishes to attend requires a strong academic record and a strong social service record for admission to its undergraduate program. Michael was very good in academics. Therefore, the University extended an offer of admission to Michael for undergraduate studies.

Which one of the following, if true, strongly supports the argument?

A) Michael got several awards for performing social service to the disabled.
B) Other applicants to the undergraduate program also performed social service.

WEAKEN

In this section on "Weaken", you will develop the ability to weaken an argument.

A complete argument is given (premises and conclusion) and your task is to pick a premise that weakens the argument.

After the argument is presented, you will be posed questions such as the following:

*Which one of the following weakens the argument the most?
*Which one of the following questions the claim of the argument?
*Which one of the following undermines the argument the most?

The correct answer is the one that weakens the argument the most. That is, if two answer choices both weaken the argument, then the one that weakens the most is the correct answer.

Incorrect answers strengthen the argument instead of weakening it.

Name _____ Date _____

| 1 | WEAKEN | Car |

The red color is associated with aggressiveness. So, a red car is very likely to get into an accident than a blue car.

Which one of the following weakens the argument the most?

A) People who wear red shirts get into more arguments than people who wear blue shirts.
B) Red motor bikes are not involved in more accidents than blue motor bikes.

| 2 | WEAKEN | Botany |

Grafting is the process of cutting two plants and joining them together. One plant is called the stock and the other plant that is joined to it is called the scion. Grafting is done to transfer good features of the scion to the stock. So, when a stock that has a strong stem is grafted with a scion that has beautiful flowers, we get a plant that has both a strong stem as well as beautiful flowers.

Which one of the following weakens the argument the most?

A) The grafting process has a very good chance of success.
B) The grafting process has only a slight chance of success.

Name _____   Date _____

3	WEAKEN	Fireworks

Fireworks triggered by computers create complex patterns that are a fantastic sight to watch. In contrast, the fireworks set off manually by hand produce only simple patterns. So, the manually lighted fireworks must be banned.

Which one of the following, weakens the argument the most?

A) Manually lighted fireworks never fail to deliver their patterns.
B) Manually lighted fireworks sometime produce bizarre patterns.

4	WEAKEN	Skin

Dandruff is a condition of the scalp where the skin becomes very flaky causing a lot of irritation. This flakiness can be caused by dry skin or by head lice, a parasitic organism. There are several varieties of shampoos available to treat dandruff. Therefore, any one of these shampoos is effective in treating dandruff.

Which one of the following weakens the argument the most?

A) The medication in the shampoos can treat both dry skin and head lice.
B) The medication in the shampoos can treat head lice only.

Name _____   Date _____

| 5 | WEAKEN | Park |

More trees in a park will attract more people to the park. This is because of the fact that the shade provided by trees is very helpful for children and adults. So, all the parks must be planted with plenty of trees.

Which one of the following weakens the argument the most?

A) Children like to play under the trees.
B) Planting trees is useless if they do not branch out wide enough.

| 6 | WEAKEN | School |

The intersection of Preston and Park roads is very dangerous. Several accidents happen at this intersection in the morning around 8 AM. It is during this time that several schools in this area open. Therefore, it is clear that the school timing is the reason for the spate of accidents at this intersection.

Which one of the following undermines the argument the most?

A) School timing is the reason for accidents in other intersections as well.
B) There have been many dangerous accidents in this intersection around 8 AM even before any school was built in this area.

Name _____   Date _____

| 7 | WEAKEN | Alarm |

Daniel lives in a neighborhood that is not free of crime. So, he secured the front and back doors with a burglar alarm. Opening any of the secured doors when the alarm is set will cause the alarm to sound. Since the alarm did not sound yesterday when his home was burgled, Daniel concluded that there is a defect in the burglar alarm.

Which one of the following undermines the argument the most?

A) The company that makes the burglar alarm that Daniel uses has warned recently that it is not one hundred percent reliable.
B) The burglar broke into Daniel's home through a rear window.

| 8 | WEAKEN | Reading |

All the friends of Jasmine read one book in addition to the prescribed book for each subject. This strategy has fetched them better grades in each subject. Therefore, Jasmine also must start reading one extra book in each subject.

Which one of the following undermines the argument the most?

A) Jasmine always gets the best grade in each subject.
B) Jasmine can also get better grades if she reads one extra book in each subject.

Name _____   Date _____

| 9 | WEAKEN | Bank |

Bank Manager: Recently, several clerks left their jobs. So, we currently do not have sufficient number of employees to provide prompt service to our customers. This should explain why there is a long queue in the bank these days.

Which one of the following undermines the Bank Manager's argument the most?

A) There was a long queue in the bank even before the clerks left their jobs.
B) It takes a lot of work to service the needs of bank customers.

| 10 | WEAKEN | Astronomy |

Astronomer: When I was star gazing last night, I spotted twenty shooting stars. So, we can expect some of these stars to fall on Earth in the next few days.

Which one of the following undermines the Astronomer's argument the most?

A) Some shooting stars have the energy to reach the Earth.
B) Shooting stars disintegrate before they reach the surface of Earth.

Name _____ Date _____

| 11 | WEAKEN | Restaurant |

Restaurant manager: The lettuce in our salad is the best that anyone can ever get in this area. We get it directly from a lettuce farmer who washes it before supplying it to us. We wash it again in our restaurant thoroughly before using it. Therefore, there is no chance for contamination of our salad.

Which one of the following undermines the argument the most?

A) Unwashed lettuce is the common reason for contamination of salads.
B) Bad tomatoes can contaminate the salad even if the lettuce is washed thoroughly.

| 12 | WEAKEN | Haircut |

Barber: Your hair is very curly. Most people do not have this much curly hair. It will take more time and effort for me to cut your hair. So, I have to charge you extra money for cutting your hair.

Which one of the following undermines the argument the most?

A) A more expensive comb must be used to untangle curly hair.
B) Hair clipping machines can cut curly hair as quickly as they can cut straight hair.

Verbal Reasoning
© Gift Of Logic, Inc * Copying prohibited

Name _____ Date _____

| 13 | WEAKEN | Car |

Car mechanic: Your car's engine must be replaced. The oxygen sensor in the engine that controls the amount of air that flows into the engine is not working. This causes an incorrect amount of air and fuel to flow into the engine.

Which one of the following undermines the argument the most?

A) Abnormal amounts of fuel and air will damage the engine.
B) The oxygen sensor alone can be replaced with a new one.
C) When one part of a machine malfunctions, the entire machine must be replaced.

| 14 | WEAKEN | Flight |

When flight# 30 left runway 22L at 9:00 AM, the pilot did not report any problems with takeoff. But, when flight# 16 left the same runway at 9:02 AM, the pilot reported problems and made an emergency request to land. Thus, it is evident that flight# 30 is responsible for the emergency landing of flight# 16.

Which one of the following undermines the argument the most?

A) The vortex of air created behind an aircraft can destabilize another aircraft that comes too close to it.
B) The pilot of flight# 16 reported a problem in one of its engines.

15	WEAKEN	Pollution

Polluted water was proven to be the cause of E-Coli bacteria in all the spinach grown in a farm. By the time this connection was found, several people who consumed this spinach were taken to hospitals with kidney problems. Therefore, produce from the farm must not be purchased.

Which one of the following undermines the argument the most?

A) Water to the entire farm is supplied by only one lake.
B) People who ate cabbage grown in the farm did not have any health related problem.

16	WEAKEN	Medical

Hypertension is caused by high blood pressure. The arteries of people with hypertension are constricted – so the heart works harder and faster to pump blood through them. Excessive amount of salt in kidneys is likely to increase the volume of blood, thereby causing high blood pressure. For this reason, only diets that are low in salt content are suitable for reducing hypertension.

Which one of the following undermines the argument the most?

A) The sodium in the salt plays an important role in causing high blood pressure.
B) Some low-salt diets are not effective in reducing hypertension.

Name _____   Date _____

ANALYTICAL REASONING

1 SUDOKU

Solve the following Sudoku. A correctly solved Sudoku has numbers 1-9 appearing only once in each row, each column and each 3x3 grid. Solving Sudokus will help you to gain valuable analytic skills.

5	8	7	4	1	2	3	9	6
6		1	5		9		7	2
2	3		8	7	6	5	1	4
8	7	5		9	3	6		1
1		3	6		8	7	5	9
9	6	4	7	5	1	2	8	3
4	1	2	3	8	7	9	6	
3				4			2	7
7	9	6	1	2	5	4	3	8

Analytical Reasoning   Answers-91

Name _____   Date _____

2
SUDOKU

Solve the following Sudoku. A correctly solved Sudoku has numbers 1-9 appearing only once in each row, each column and each 3x3 grid. Solving Sudokus will help you to gain valuable analytic skills.

3	9	7	8	4	5	1	6	2
2		5	6		9	7		3
4	1	6	7	3	2	8	5	9
6		4		8	1		7	5
1	5		2			6		8
7		8		6	3		9	1
8		3		5	7		2	6
9	7		3		6	5		
	6	2	4	9		3	1	7

Analytical Reasoning

Name _____   Date _____

3
SUDOKU

Solve the following Sudoku. A correctly solved Sudoku has numbers 1-9 appearing only once in each row, each column and each 3x3 grid. Solving Sudokus will help you to gain valuable analytic skills.

4	6		5	2	1	7		9
7		2	4		8		5	1
5	8	1	9		6	4	2	3
1	3	5	8	6	7		4	2
	4	6	3	9	2	1	7	5
9	2	7	1		4	8		6
6		8		4	5	3	9	7
2	7	9	6	8		5	1	4
3	5	4	7	1	9		6	8

Analytical Reasoning
© Gift Of Logic, Inc * Copying prohibited

Name _____ Date _____

4
SUDOKU

Solve the following Sudoku. A correctly solved Sudoku has numbers 1-9 appearing only once in each row, each column and each 3x3 grid. Solving Sudokus will help you to gain valuable analytic skills.

9		7	1		8	6		3
3	6		7	5	2	8		9
2		4	9	6	3		7	5
1		9	5		4	3		6
5	4		6	8		7	9	
6		8		3	9		1	4
7		2	3		5	4		8
4	1		8	2		9	3	
	3	6		9	7		5	1

Analytical Reasoning Answers-94
© Gift Of Logic, Inc * Copying prohibited

Name _____   Date _____

5
SUDOKU

Solve the following Sudoku. A correctly solved Sudoku has numbers 1-9 appearing only once in each row, each column and each 3x3 grid. Solving Sudokus will help you to gain valuable analytic skills.

5	2	1	9	3	6		8	
	7	6		5	1	9	2	3
3	9	4	8	2	7	5	6	1
4		7	2		9	3	5	
1	8	9	5	6	3	2	4	7
	3			4			1	
9	1		6	8		4	7	
	5	8		7	4		9	2
	4	2	1		5	8		6

Analytical Reasoning

1 POSITIONING — immediately

SCENARIO

Tom, Dick, and Harry have to be seated in three chairs, numbered 1, 2 and 3 and placed next to each other. Tom must sit immediately to the left of Harry.

QUESTIONS

1) Can Tom sit in the third chair?
   A) Yes   B) No

2) Can Tom sit in the first chair?
   A) Yes   B) No

3) Can Harry sit in the first chair?
   A) Yes   B) No

4) Can Dick sit in the third chair?
   A) Yes   B) No

5) Dick can sit in the first chair or the third chair.
   A) True   B) False

Analytical Reasoning
© Gift Of Logic, Inc * Copying prohibited

| Name _____ | Date _____ |

2 POSITIONING   next to

SCENARIO

Rolly, Polly, and Molly have to sleep in three beds placed next to each other. The beds are named bed# 1, bed# 2 and bed# 3. Rolly and Polly must sleep next to each other.

QUESTIONS

1) Can Molly sleep in bed# 2?
    A) Yes   B) No

2) Can Polly sleep in bed# 1?
    A) Yes   B) No

3) If Molly sleeps in bed# 1, where can Rolly sleep?
    A) bed# 2 only   B) bed# 2 or bed# 3

Name _____   Date _____

| 3 | POSITIONING | fixed position |

SCENARIO

Vivek, Tom, Sandra, and Nazia are to be photographed for a group picture and must stand in four spots next to each other.

Vivek must stand in the first spot. Sandra must stand in the third spot.

QUESTIONS

1) Can a picture be taken in the following order?
   Vivek, Sandra, Tom, Nazia

   A) Yes   B) No

2) Write the possible positions of the four people in which a picture can be taken.

3) How many group pictures can be taken?
   A) 1   B) 2

Name _____   Date _____

| 4 | POSITIONING | before |

SCENARIO

Three cars - a Toyota, a Honda, and a Ford must be parked in parking spots numbered 1, 2, and 3 that are next to each other in a line.

The Toyota must be parked before the Honda.

QUESTIONS

1) Can Toyota be in the third spot?
   A) Yes   B) No

2) Can Toyota be in the second spot?
   A) Yes   B) No

3) Can Ford be in the first spot?
   A) Yes   B) No

4) Can Toyota be in the first spot?
   A) Yes   B) No

5) Can Honda be in the first spot?
   A) Yes   B) No

Name _____ Date _____

5   POSITIONING   after

SCENARIO

A green, a black, and a red shoe must be arranged in a shoe rack one after the other. The black shoe must be placed after the red shoe.

QUESTIONS

1) If the green shoe is placed in the third position, then the black shoe must be placed in which position?
     A) 1   B) 2

2) If the green shoe is placed in the second position, then the black shoe must be placed in which position?

     A) 1   B) 3

3) If the green shoe is placed in the first position, then the black shoe must be placed in which position?
     A) 2   B) 3

Analytical Reasoning   Answers-102
© Gift Of Logic, Inc * Copying prohibited

6	POSITIONING	immediately before

SCENARIO

Three boats are to sail in the Mississippi river one after the other in the westward direction. They are named B1, B2, and B3. B2 must sail immediately before B1.

QUESTIONS

1) If B2 sails in the middle, then B3 can sail in which position?
   A) 1   B) 3

2) If B2 sails first, then B3 can sail in which position?
   A) 2   B) 3

3) Can B2 sail in the last position?
   A) Yes   B) No

Name _____   Date _____

7   POSITIONING   fixed position if-then

SCENARIO

Three donkeys, named Abra, Babra, and Cabra have to be taken out for a walk one behind the other.

If Abra walks in the first position, then Babra must be in the third position.
If Babra is in the second position, then Cabra must be in the first position.

QUESTIONS

1) Can the donkeys walk in the following order?
   Abra, Babra, Cabra

   A) Yes   B) No

2) Can the donkeys walk in the following order?
   Cabra, Babra, Abra

   A) Yes   B) No

3) Can the donkeys walk in the following order?
   Cabra, Abra, Babra

   A) Yes   B) No

8 POSITIONING — cannot

SCENARIO

Johnny wants to wear three watches in her hand one after the other. The watches are labeled as W1, W2, and W3.

W2 cannot be the first watch.

QUESTIONS

1) Verify whether each of the following choices could represent the possible positions of the watches that Johnny can wear from first to last.

   A) W2,W1,W3
   B) W3,W1,W2
   C) W1,W3,W2
   D) W2,W3,W1
   E) W3,W2,W1
   F) W1,W2,W3

2) If W3 also should not be worn first, then write the possible ways in which the watches can be worn.

Name ——————————————— Date ———————————————

| 9 | POSITIONING | or, cannot |

SCENARIO

Three books, labeled B1, B2, and B3 must be placed next to each other in three spots in a bookshelf.

B1 can be in position 1 or in position 2, but it cannot be in position 3.
B3 must not be in position 2.

QUESTIONS

1) Is the following ordering of books correct?
   B2, B1, B3
   A) Yes   B) No

2) Is the following ordering of books correct?
   B1, B3, B2
   A) Yes   B) No

3) Can the books be placed in the following order?
   B3, B2, B1
   A) Yes   B) No

10

POSITIONING — or, consecutive, possibilities

SCENARIO

Abigail must wear a blue shirt or a green shirt to school everyday. She cannot wear the same colored shirt on consecutive days.

QUESTIONS

1) Can Abigail wear shirts in the order shown for each week? Write Yes/No in each row in the column captioned as 'Possible'.

Week	Monday	Tuesday	Wednesday	Thursday	Friday	Possible?
1	Blue	Green	Blue	Blue	Green	
2	Blue	Green	Blue	Green	Blue	
3	Green	Blue	Green	Green	Blue	
4	Green	Blue	Green	Blue	Green	

11

SCENARIO

Amber must wear a blue skirt or a green skirt to school everyday. She can wear the same colored skirt on consecutive days.

QUESTIONS

1) Can Amber wear the skirts in the order shown in the table below? Write Yes/No in each row in the column captioned as 'Possible'.

Week	Monday	Tuesday	Wednesday	Thursday	Friday	Possible?
1	Blue	Green	Blue	Green	Blue	
2	Blue	Blue	Blue	Blue	Red	
3	Green	Green	Blue	Blue	Green	

Analytical Reasoning

| **12** | POSITIONING | or, possibilities |

SCENARIO

Kate can wear a blue or a green skirt on Mondays, an Orange or a Yellow skirt on Tuesdays, and a Red or a Green skirt on Wednesdays.

QUESTIONS

1) Which of the following choices are possible? Write Yes/No in the column marked "Possible" for each of the choices shown.

Monday	Tuesday	Wednesday	Possible?
Orange	Yellow	Green	
Green	Orange	Green	
Blue	Green	Orange	
Green	Yellow	Green	

Analytical Reasoning
© Gift Of Logic, Inc * Copying prohibited

13 POSITIONING and/or combined possibilities

SCENARIO

Mike plays soccer and hockey on Wednesdays, hockey or baseball on Thursdays, and skates or swims, but not both on Fridays.

QUESTIONS

1) Which of the following scenarios are possible?

Wednesday	Thursday	Friday	Possible?
Hockey and Baseball	Soccer	Skating	
Soccer and Hockey	Baseball	Skating	
Skating and Swimming	Hockey	Soccer	
Soccer and Hockey	Hockey and Baseball	Skating and Swimming	
Hockey and Soccer	Baseball	Swimming	

Analytical Reasoning   Answers-109

GROUPING

1 selection/combination

A box contains the following three objects: paper, pencil, and eraser

Write below all the possible ways in which you could pick any two objects from the box.

---

2 selection/ combination

Billy can choose two of the following animals as pets: cat; dog; mouse

Write all the possible combinations of animals that Billy could choose as pets.

GROUPING

3

select two with condition

Rick went to the furniture store to purchase any two from the following list: sofa; table; mirror; chair

Sofa and chair must not be purchased together.

1) Which of the following purchases can be made?
   A) sofa, table
   B) table, chair
   C) chair, sofa
   D) chair, mirror

---

4 select two with condition

   Toyota   Honda   Mazda

Select two cars from the above list.
Toyota and Mazda must be selected together.

1) Which of the following selections can be made?

   A)   Toyota, Honda
   B)   Honda, Mazda
   C)   Mazda, Toyota

Name ——————————— Date ———————

GROUPING

5 two bunches

Carmen bought a pack of six balloons for her birthday party - 2 blue, 2 green and 2 red. She must tie all the six balloons in two bunches. The two red balloons must be in the same bunch.

1) Which of the following choices have valid bunches?

    A) Bunch-1: Blue, Blue, Green   Bunch-2: Green, Red, Red
    B) Bunch-1: Blue, Red, Red   Bunch-2: Green, Blue, Blue
    C) Bunch-1: Blue, Red, Red   Bunch-2: Green, Blue, Green

---

6 three bunches

From a pack of six balloons (1 red, 2 blue and 3 green), three bunches are to be made, 2 balloons in each bunch. A green balloon must be in each bunch.

1) Which of the following choices have valid bunches?

	Bunch-1	Bunch-2	Bunch-3
A)	Green, Red	Blue, Blue	Green, Green
B)	Green, Red	Green, Blue	Green, Blue
C)	Blue, Blue	Green, Red	Green, Green

Analytical Reasoning
© Gift Of Logic, Inc * Copying prohibited

Name —————————————— Date ——————————————

GROUPING

7 combinations

Four balls on the table are marked as follows:
   Red1, Red2, Green3, Green4

1) Pick one red ball and one green ball from the table. Write all the possible ways in which the balls can be selected.

2) Pick two red balls and one green ball from the table. Write all the possible ways in which the balls can be selected.

3) Pick two green balls and one red ball. Write all the possible ways in which the balls can be selected.

Name ——————————————— Date ———————————————

GROUPING

8 condition

Four flowers have the following colors: Red, Green, Blue, and Yellow. Select two flowers. The red and blue flowers must not be selected together. The green and yellow flowers must not be selected together.

1) Which of the following selections are correct?
   A) Red, Blue
   B) Blue, Red
   C) Red, Yellow
   D) Yellow, Green

---

9 condition

Four flowers have the following colors: Red, Green, Blue, and Yellow. Select three flowers. The red and the blue flower must not be selected together. The red and yellow flower must not be selected together.

1) Which of the following selections are correct?
   A) Red, Green, Yellow
   B) Red, Green, Blue
   C) Red, Yellow, Blue
   D) Green, Blue, Yellow

Analytical Reasoning
© Gift Of Logic, Inc * Copying prohibited

Name _____   Date _____

GROUPING

10 no conditions

Following are the contents of two boxes.
  Box#1: apple1, apple2, apple3
  Box#2: orange1, orange2, orange3, orange4

Select two apples and two oranges from the two boxes.

1) Which of the following is a correct selection?

A) apple1, apple2, orange3, apple3
B) apple1, apple3, orange1, orange2

---

11 condition

Following are the contents of two boxes.
  Box#1: apple1, apple2, orange3, orange4
  Box#2: orange1, orange2, apple3, apple4

Select two apples and two oranges.
Fruits of the same kind must not be selected from the same box.

Which of the following is a correct selection?
  A) apple1, apple2, orange3, orange1
  B) apple1, apple3, orange3, orange4
  C) apple1, apple2, orange1, orange2
  D) apple1, apple4, orange2, orange3

Analytical Reasoning   Answers-115
© Gift Of Logic, Inc * Copying prohibited

Name _____   Date _____

GROUPING

12 pick a few

A sports utility box has the following objects in it.

baseball bat
ping pong ball
tennis ball
ping pong bat
soccer ball
golf ball

Pick two bats and three balls and write your selection below.

---

13 at least

parrot, lion, cat, butterfly, tiger, horse, eagle, dove

From the list above, select two birds and at least two animals. Write your selection below.

Analytical Reasoning   Answers-115
© Gift Of Logic, Inc * Copying prohibited

Name ——————————————— Date ———————————

GROUPING

14 no more, no less

car1, truck1, car2, truck2, truck3, car3, truck4

Select no more than two cars and no less than three trucks from the list shown above.

1) What is the least number of cars that can be picked?

2) What is the most number of cars that can be picked?

3) What is the least number of trucks that can be picked?

4) What is the most number of trucks that can be picked?

---

15 at most

triangle1, circle1, triangle2, circle2, triangle3, circle3

Select at most two circles and at most three triangles from the list shown above.

1) What is the least number of circles that can be picked?

2) What is the most number of circles that can be picked?

3) What is the least number of triangles that can be picked?

4) What is the most number of triangles that can be picked?

Name _____   Date_____

PICTORIAL REASONING

Name _____ Date _____

PATTERN PERCEPTION - MISSING PATTERN

Find the correct figure from the three alternatives given that will fit logically into the missing portion of the figure on the left.

1 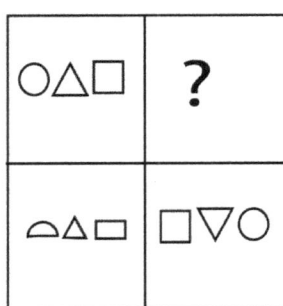 A  B  C

2 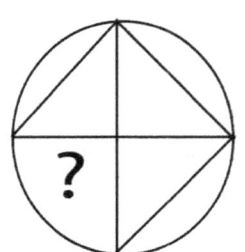 A  B 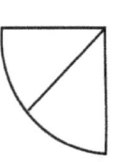 C

3 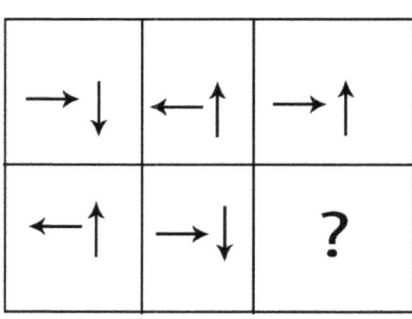 A 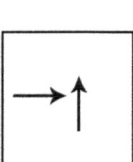 B 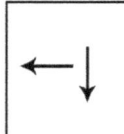 C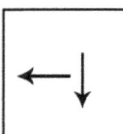

4 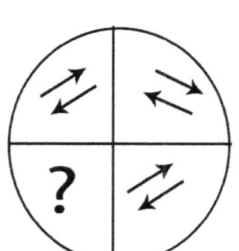 A  B  C

Pictorial Reasoning Answers-117 52
© Gift Of Logic, Inc * Copying prohibited

Name _____ Date _____

PATTERN PERCEPTION - CONTINUING PATTERN

Find the correct figure from the two alternatives given, that will logically continue the pattern of figures on the left.

5 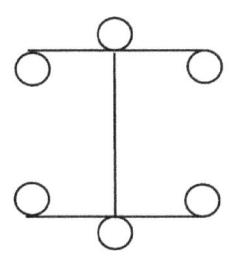 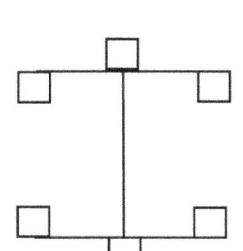 ?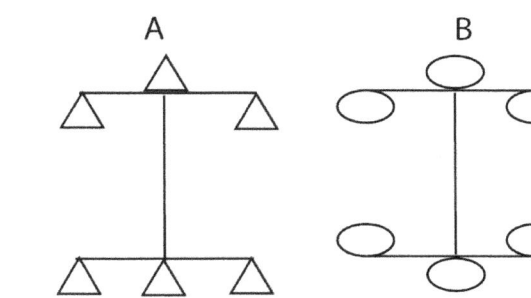

6 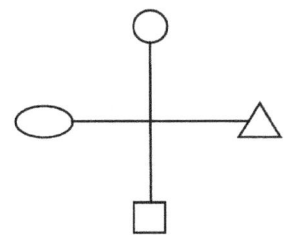 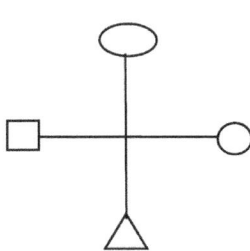 ?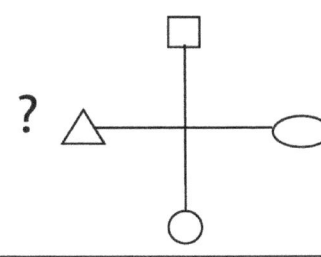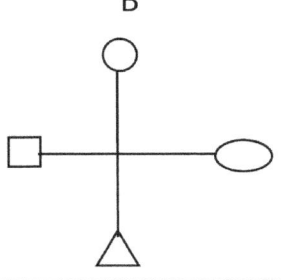

7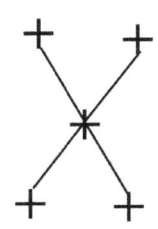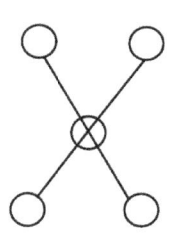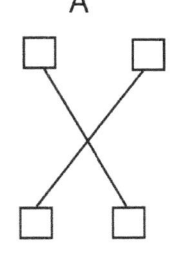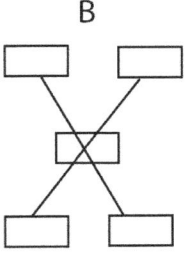

8 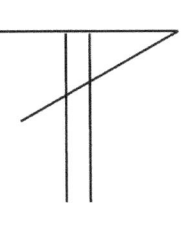 ?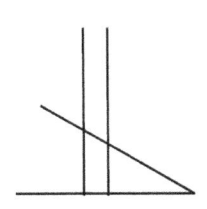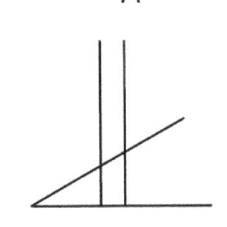

Pictorial Reasoning
© Gift Of Logic, Inc * Copying prohibited

Name _____   Date _____

FIGURE FORMATION

Find the correct figure that will combine with the figure on the left to form the figure on the right.

1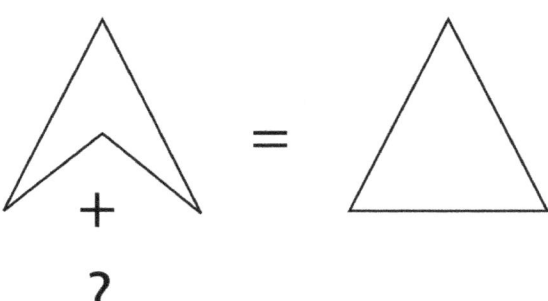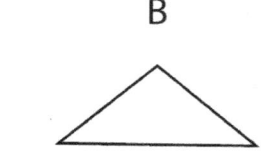

2

3

4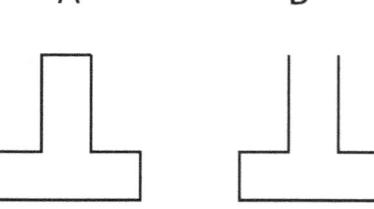

Pictorial Reasoning   Answers-117   54
© Gift Of Logic, Inc * Copying prohibited

Name _____   Date _____

FIGURE FORMATION

Find the correct figure that will be formed when the figures on the left are combined together.

5 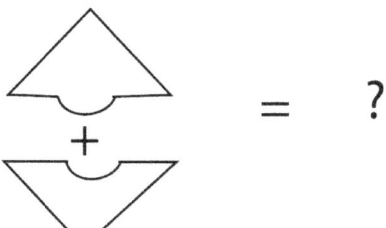 = ?   A 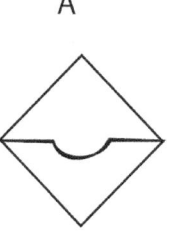   B

6 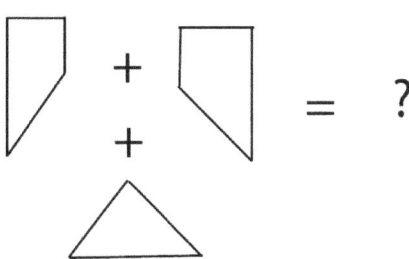 = ?   A 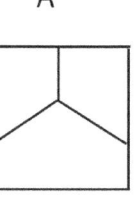   B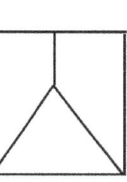

7 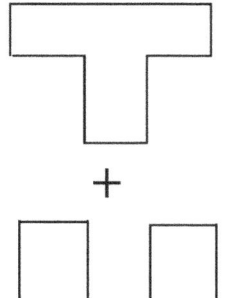 = ?   A 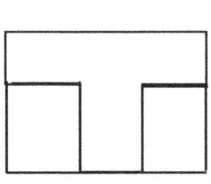   B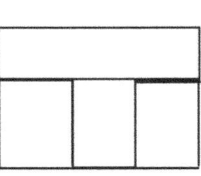

8 = ?   A 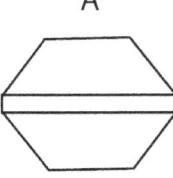   B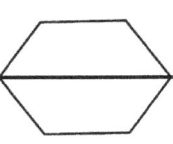

Pictorial Reasoning   Answers-117
© Gift Of Logic, Inc * Copying prohibited

Name _____   Date _____

PAPER FOLDING AND CUTTING

Find the correct figure that will be formed when the paper on the left is folded in the direction of the arrows, and then holes are cut in it as shown.

1   A   B   C   D

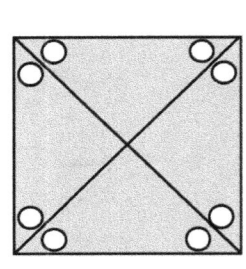 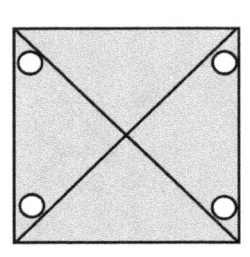 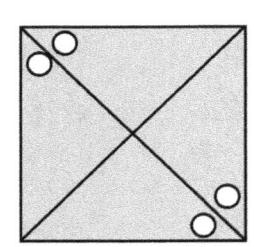 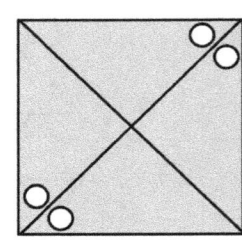

2   A   B   C   D

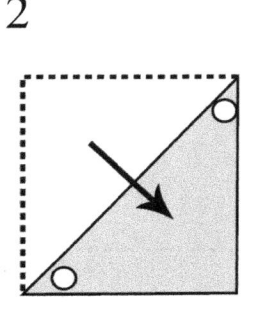 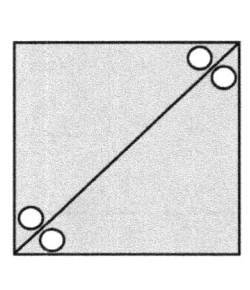 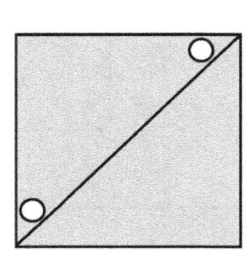 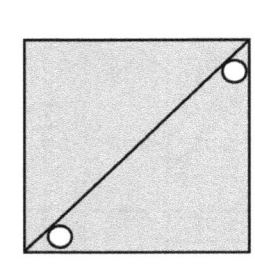 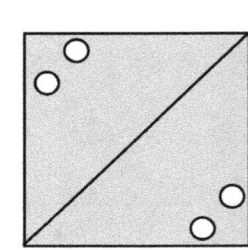

3   A   B   C   D

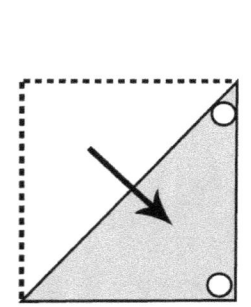 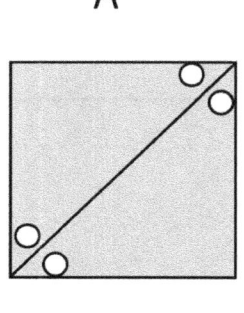 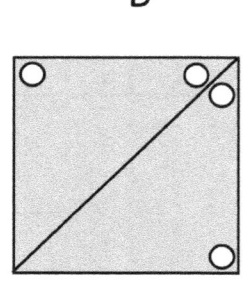 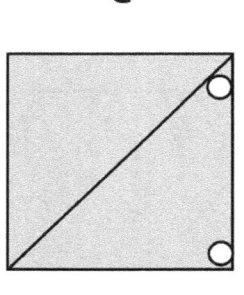 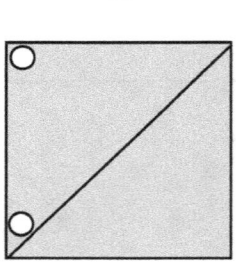

Pictorial Reasoning   Answers-118
© Gift Of Logic, Inc * Copying prohibited

Name ——————————————— Date ———————————

FIGURE MATRIX- ANALOGY

Find the correct figure from the alternatives given that will fit in the empty box, such that the bottom two figures are related in the same way as the top two figures.

1      A    B

2      A    B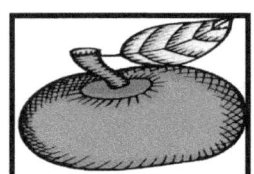

3      A 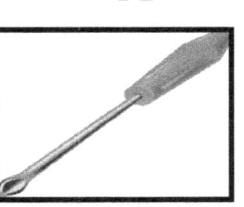   B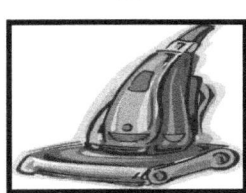

4      A    B

Pictorial Reasoning
© Gift Of Logic, Inc * Copying prohibited

Name _____   Date _____

FIGURE MATRIX- SIMILARITY

Three figures in the 2 x 2 matrix have similar characteristics. Find the fourth figure from the alternatives given that is also alike.

5.

6.

7.

8.

Pictorial Reasoning   Answers-118   58
© Gift Of Logic, Inc * Copying prohibited

Name _____ Date _____

RULE DETECTION

Read the rule given rule in each question. Then, find the correct choice from the alternatives given that satisfies the rule.

1. Objects grow as the sequence progresses

A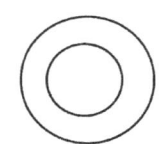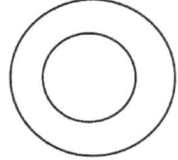

B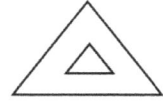

2. Objects split as the sequence progresses

A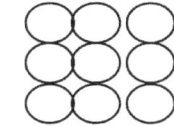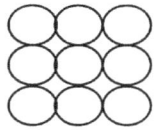

B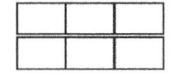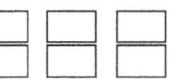

3. Objects join as the sequence progresses

A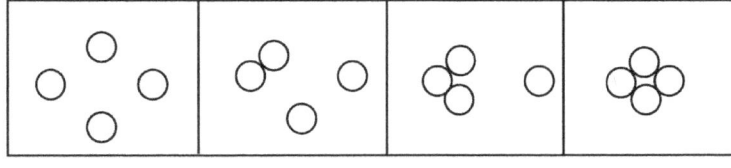

B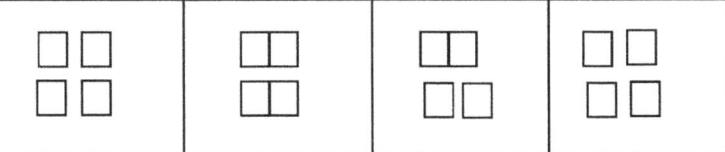

Pictorial Reasoning

Name —————————————— Date ——————————

RULE DETECTION

Read the given rule in each question. Then, find the correct choice from the alternatives given that satisfies the rule.

Figures close progressively

4

A

B

C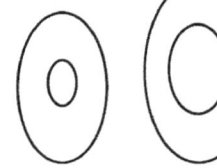

Inside and outside figures have the same center

5

A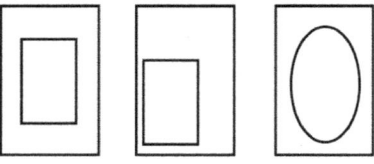

B

Intersecting area increases as the sequence progresses

6

A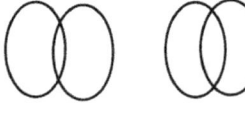

B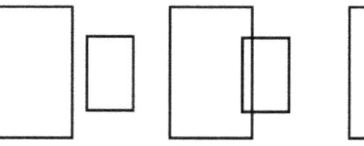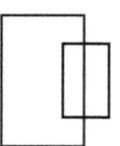

Pictorial Reasoning

ANSWERS

| 1 | STRENGTHEN |

Exactly one hundred days ago..

Which one of the following, if true, most strengthens the argument?
A) On that day, there were many dark clouds in the sky.
B) On that day, there was bright sunshine.

ANSWER

Answer: A

A – correct – this premise strengthens the argument. Add this as a premise to the argument and see if it strengthens or weakens the argument.

> 100 days ago, temperature was very cold.
> On that day, there were many dark clouds in the sky.
> Therefore, it rained that day.

The fact that there were dark clouds in the sky only strengthens the conclusion that it rained that day.

B – incorrect. Add this as a premise to the argument and see if it strengthens or weakens the argument.
> 100 days ago, temperature was very cold.
> On that day, there was bright sunshine.
> Therefore, it rained that day.

Bright sunshine is not normal during a rainy day. So, this choice does not strengthen the argument.

Answers

| 2 | STRENGTHEN |

Tomatoes in several plants in Joe's garden..

Which one of the following, if true, most strengthens the argument?
A) Joe's sister Amanda saw rabbits chewing the tomatoes.
B) Joe's garden is protected by an alarm to scare the rabbits.

ANSWER

Answer: A

A – correct – this premise reinforces Joe's conclusion. Add this premise to the other premises and you will see that this choice strengthens the argument.

Tomatoes were found in half-bitten state.
Joe knew that there were rabbits in the garden.
Joe's sister Amanda saw rabbits chewing the tomatoes.
So, Joe was sure the rabbits were responsible.

B – incorrect – this choice weakens the argument. If there is an alarm to scare the rabbits, then there is less likelihood of rabbits being able to chew the tomatoes. So, this premise actually weakens the argument.

Tomatoes were found in half-bitten state.
Joe knew that there were rabbits in the garden.
Joe's garden is protected by an alarm to scare the rabbits.
So, Joe was sure the rabbits were responsible.

Clearly, the inclusion of this premise weakens the argument.

3 STRENGTHEN

Laura is the person responsible for arranging..

Which one of the following, if true, most strengthens the argument?
A) The hotel requires at least five people to attend a reunion.
B) Holding a reunion will make sense only if all the family members attend.

ANSWER

Answer: B

A – incorrect – put all the premises and conclusion together.
   Only ten members can attend.
   The hotel requires at least five people to attend a reunion.
   Therefore, Laura decided to cancel the reunion.

If ten members can attend and the hotel requires only five, then why did Laura decide to cancel? So, this choice does not strengthen the conclusion.

B – correct – put the premises and conclusion together.
   Only ten members can attend.
   Reunion will make sense only if all the family members can attend.
   Therefore, Laura decided to cancel.

From the first premise - "only ten members can attend", it appears that several others were not able to attend. From the second premise, it appears that holding the reunion will not make sense in this case. So, this choice strengthens Laura's decision to cancel the reunion.

Answers
© Gift Of Logic, Inc * Copying prohibited

| 4 | STRENGTHEN |

Seventy percent of the road accidents..

Which one of the following, if true, strongly supports the argument?
A) Fines do not have an influence on driver behavior.
B) Drivers do not like to pay fines.

| ANSWER |

Answer: B

A - incorrect - add this fact to the argument and analyze.
   Seventy percent of accidents are caused by speeding drivers.
   Thirty percent of the accidents occur for other reasons.
   Fines do not have an influence on driver behavior.
   Therefore, we must increase the fines on speeding drivers.

Obviously, if fines do not have an influence on driver behavior, then, increasing the fines (conclusion) is not going to prevent a majority of accidents. So, this choice weakens the conclusion.

B - correct - add this fact to the argument and analyze.
   Seventy percent of accidents are caused by speeding drivers.
   Thirty percent of the accidents occur for other reasons.
   Drivers do not like to pay fines.
   Therefore, we must increase the fines on speeding drivers.

This choice says that drivers do not like to pay fines. So, it adds strength to the conclusion that we must increase the fines for over speeding, hoping that it will influence the drivers to not speed over the limits.

5	STRENGTHEN

Farmer Joe noted nothing unusual in his small..

Which one of the following, if true, most strengthens the argument?
A) Earth in farmland can crack suddenly due to insufficient irrigation.
B) Seismographs in the area recorded an earthquake last night.

ANSWER

Answer: B

A – incorrect – add this choice to the argument and analyze.

> Trembling movement occurred last night.
> Earth in farmland can crack suddenly due to insufficient irrigation.
> So, the earthquake of previous night caused the crack.

Obviously, this choice does not support the conclusion that the earthquake caused the crack. It points to a different cause, namely insufficient irrigation, for the crack in the farmland.

B – correct – add this choice to the argument and analyze.

> Trembling movement occurred last night.
> Seismographs in the area recorded an earthquake last night.
> So, the earthquake of previous night caused the crack.

This choice provides additional proof that there was an earthquake last night. So, this choice strengthens the conclusion.

6 STRENGTHEN

Farmers growing apples bring them to the market..

Which one of the following, if true, most strengthens the argument?
A) Food inspectors tested the apples and found worms inside.
B) Food inspectors tested the apples and did not find worms inside.

ANSWER

Answer: B

A – incorrect – add this premise to the argument and analyze.
    Last year, high demand for apples was met.
    Reports of worms inside apples this year.
    High demand for apples was met this year.
    Food inspectors tested the apples and found worms inside.
    Therefore, the apples brought to the market this year are safe to eat.

This choice presents a fact that weakens the conclusion. If worms were found in the apples, then it cannot support the conclusion that it is safe to eat the apples.

B – correct - add this premise to the argument and analyze.
    Last year, high demand for apples was met.
    Reports of worms inside apples this year.
    High demand for apples was met this year.
    Food inspectors tested the apples and did not find worms inside.
    Therefore, the apples brought to the market this year are safe to eat.

Since the food inspectors did not find any worms in the apples, this choice strengthens the conclusion that the apples are safe to eat this year.

7 STRENGTHEN

If an actor is wealthy, he will donate..

Which one of the following, if true, most strengthens the argument?
A) If an actor donates, he is wealthy.
B) If an actor donates, he will be popular.

ANSWER

Answer: C
The given argument is represented in symbolic as follows:
   wealthy → donate
   therefore, wealthy → popular

A - incorrect - add the premise to the argument and analyze.
   wealthy → donate
   donate → wealthy
   therefore, wealthy → popular

This choice does not tie wealth or donations to popularity and so it does not strengthen the argument.

B - correct - add the premise to the argument and analyze.
   wealthy → donate
   donate → popular
   therefore, wealthy → popular (inferred from conditional chaining principles - refer to the book titled "Primer" for discussion on conditional reasoning). This choice strengthens the argument by tying donations and popularity and is the correct answer.

Answers
© Gift Of Logic, Inc * Copying prohibited

8	STRENGTHEN

Soccer is a game that requires..
Which one of the following, if true, most strengthens the argument?
A) Drink-A and Drink-B have the same ingredients.
B) Drink-A is specially formulated for males and Drink-B is specially formulated for females.
C) Male dogs tire out sooner than female dogs when they consume Drink-B and female dogs tire out sooner than male dogs when they consume Drink-A.

ANSWER

Answer: B
The argument is shown below:
   Girls tire sooner than boys with Drink-A.
   Boys tire sooner than girls with Drink-B.
   So, the company sells Drink-A for boys and Drink-B for girls.

Add the choices to the argument and analyze.

A - incorrect - if both the drinks have the same ingredients, this does not strengthen the company's decision to specifically sell Drink-A for boys and Drink-B for girls.

B - correct - this choice gives additional information about the drinks that helps to strengthen the argument that Drink-A is for boys and Drink-B is for girls.

C - incorrect - even though this premise seems to strengthen the argument, it is related to dogs and not boys and girls.

9 STRENGTHEN

The coach of a defeated team tried to figure out..

Which one of the following, if true, strongly supports the argument?
A) The winning team showed better team spirit during the game.
B) The defeated team members argued about their roles during the game.

ANSWER

Answer: B

A - incorrect - add the choice to the argument and analyze.
 Insufficient practice or lack of team spirit may be the reasons for the loss.
 Lack of coordination was seen during the game.
 The winning team showed better team spirit during the game.
 Therefore, introduce more team building activities.

This choice strengthens the argument to some extent, but not as strongly as choice B. The coach is analyzing the defeat of his team, not the other.

B - correct - add the choice to the argument and analyze.
 Insufficient practice or lack of team spirit may be the reasons for the loss.
 Lack of coordination was seen during the game.
 The defeated team members argued about their roles during the game.
 Therefore, introduce more team building activities.

The defeated team members argued about their roles during the game. This would have led to a lack of team spirit. Hence, the coach is justified in introducing more team building activities.

10 STRENGTHEN

Brendan had the option of buying the blue car..

Which one of the following, if true, strongly supports the argument?
A) The red car is known to be less problematic than the blue car.
B) The blue car consumes less fuel than the red car.

ANSWER

Answer: A

Note that there are two conclusions in the argument. Both the conclusions start with "So". The argument can be described shortly as follows:

> Brendan can buy the blue car or the red car.
> Blue car was cheaper than the red car.
> So, Brendan bought the blue car.
> Blue car broke down several times, expensive repairs.
> So, Brendan decided to sell the blue car and buy the red car.

A - correct - since the red car is known to be less problematic than the blue car, this choice strengthens the argument's conclusion.

B - incorrect - this choice strengthens the first conclusion, that is, Brendan's decision to buy the blue car. But, this does not strengthen the second conclusion - Brendan's decision to sell his blue car. The second conclusion is the overriding conclusion. This conclusion must be strengthened, but this choice does not do that.

Answers
© Gift Of Logic, Inc * Copying prohibited

11	STRENGTHEN

Two types of shots were offered by a medical clinic ..

Which one of the following, if true, strongly supports the argument?
A) The shots helped several people resist bronchitis and pneumonia.
B) Several people who took both the shots reported no resistance to bronchitis and pneumonia.

ANSWER

Answer: B

The given argument is shown below:
>Type-A shots to prevent bronchitis.
>Type-B shots to prevent pneumonia.
>Linus took both shots.
>Linus contracted both bronchitis and pneumonia.
>Therefore, the shots were useless.

Now, add the choices to the above argument and analyze.

A - incorrect - if several people were able to resist bronchitis and pneumonia, then this fact weakens the conclusion that the shots were useless.

B - correct - this choice indicates that several people reported no resistance to bronchitis and pneumonia. So, apart from Linus, several other people also did not respond positively to these shots. So, this choice gives one more reason to believe that the shots were useless.

12 STRENGTHEN

Psychologist: When people make small mistakes..

Which one of the following, if true, strongly supports the argument?
A) People are not afraid of reporting small mistakes.
B) People are ashamed to admit making big mistakes.

ANSWER

Answer: B

The argument is described below:
> When people make small mistakes they report it.
> When people make big mistakes they don't report it nor do they tell the truth when questioned.
> Therefore, people are not honest at all times.

Add the answer choices to the argument and analyze.

A-incorrect- if people are not afraid of reporting small mistakes, this fact does not help to strengthen the conclusion that they are not honest at all times.

B - correct - if people are ashamed to admit big mistakes, they are likely to not report it voluntarily or confess to it when questioned. So, this will strengthen the conclusion that people are not honest at all times.

| 13 | STRENGTHEN |

The University that Michael wishes to attend..

Which one of the following, if true, strongly supports the argument?
A) Michael got several awards for performing social service to the disabled.
B) Other applicants to the undergraduate program also performed social service.

ANSWER

Answer: A

> University requires strong academic and strong social service record.
> Michael was very good in academics.
> Therefore, the University offered him admission.

Now, add the answer choices to the argument and analyze.

A - correct - It is a requirement that one should have a strong academic and a strong social service record. The original argument says that Michael has a strong academic record, but does not say anything about his social service record. This choice indicates that he has a strong social service record too. Therefore, this choice strengthens the argument.

B - incorrect - the conclusion of the argument is regarding the University offering admission to Michael based on his strong academic and social service record. If other applicants also did social service, that fact does not strengthen the decision of the University to offer admission to Michael.

| 1 | WEAKEN |

The red color is associated with..

Which one of the following, weakens the argument the most?
A) People who wear red shirts get into more arguments than people who wear blue shirts.
B) Red motor bikes are not involved in more accidents than blue motor bikes.

ANSWER

Answer: A

Add the choices to the given argument and analyze whether it weakens the argument or not.

A - incorrect - this choice strengthens the argument.
    Red color is aggressive.
    People wearing red shirts get into more arguments than people wearing blue shirts.
    So, a red car is very likely to get into an accident than a blue car.

B - correct - this choice weakens the argument.
    Red color is aggressive.
    Red bikes are not involved in more accidents than blue bikes.
    So, a red car is very likely to get into an accident than a blue car.

If red bikes are not involved in more accidents than blue bikes, then this weakens the conclusion that a red car will get into an accident more likely than a blue car.

Answers

2	WEAKEN

Grafting is the process of cutting two plants..

Which one of the following, weakens the argument the most?
A) The grafting process has a very good chance of success.
B) The grafting process has only a slight chance of success.

ANSWER

Answer: B

A - incorrect - this statement strengthens the argument. Arrange the premises and conclusion together.

> Grafting transfers good features from scion to stock.
> Grafting process has a very good chance of success.
> So, strong stem grafted with beautiful flowers will yield a plant that has both.

Since the grafting process has a very good chance of success, it strengthens the conclusion that we can get beautiful flowers grafted on to the strong stem.

B - correct - this statement weakens the argument. Arrange the premises and conclusion together.

> Grafting transfers good features from scion to stock.
> Grafting process has only a slight chance of success.
> So, strong stem grafted with beautiful flowers will yield a plant that has both.

If the grafting process has only a slight chance of success, then it weakens the claim that beautiful flowers can be grafted onto a strong stem.

3 WEAKEN

Fireworks triggered by computers..

Which one of the following, weakens the argument the most?
A) Manually lighted fireworks never fail to deliver their patterns.
B) Manually lighted fireworks sometime produce bizarre patterns.

ANSWER

Answer: A -   Add the choices to the argument and analyze.

A - correct
  Fireworks set off by computers have complex patterns.
  Fireworks set off manually have simple design patterns.
  Manually lighted fireworks never fail to deliver their patterns.
  So, manually lighted fireworks must be banned.

This choice weakens the argument. If they never fail to deliver their patterns, then why should they be banned?

B - incorrect
  Fireworks set off by computer software have complex patterns.
  Fireworks set off manually have simple design patterns.
  Manually lighted fireworks sometime produce bizarre patterns.
  So, manually lighted fireworks must be banned.

This choice strengthens the argument that manual fireworks must be banned. Bizarre patterns only give additional support for banning manual fireworks.

Answers
© Gift Of Logic, Inc * Copying prohibited

4	WEAKEN

Dandruff is a condition of the scalp where..

Which one of the following, weakens the argument the most?
A) The medication in the shampoos can treat both dry skin and head lice.
B) The medication in the shampoos can treat head lice only.

ANSWER

Answer: B - Add the choices to the argument and analyze.

A - incorrect

    Dandruff is when scalp becomes flaky.
    Dry skin or head lice c→ flakiness (causal relation).
    The medication can treat both dry skin and head lice.
    So, any shampoo is effective in treating dandruff.

This choice strengthens the argument. If all the shampoos have medication in them to treat both the causes of flakiness, then they are all going to be effective in treating dandruff.

B - correct - this weakens the argument

    Dandruff is when scalp becomes flaky.
    Dry skin or head lice c→ flakiness
    The medication in the shampoos can treat head lice only.
    So, any shampoo is effective in treating dandruff.

Note carefully that head lice is one of the causes for dandruff. Dry skin is another cause. So, if the medication in the shampoos can treat head lice only, then the shampoos will not be effective in treating dandruff on a person having dry skin. So, this statement weakens the argument.

Answers
© Gift Of Logic, Inc * Copying prohibited

5	WEAKEN

More trees in a park..

Which one of the following, weakens the argument the most?
A) Children like to play under the trees.
B) Planting trees is useless if they do not branch out wide enough.

ANSWER

Answer: B

Add the choices to the argument and analyze.

A - incorrect - this choice does not weaken the conclusion. If children like to play under the trees, then this fact only strengthens the argument for planting plenty of trees in all the parks.

> More trees in a park attracts more people to the park.
> Shade provided by trees is helpful for children and adults.
> Children like to play under the trees.
> So, all the parks must be planted with plenty of trees.

B - correct - this choice weakens the argument by casting a doubt that if the trees do not branch out wide enough, then it will be useless to plant them as they will not provide the shade that will be helpful.

> More trees in a park attracts more people to the park.
> Shade provided by trees is helpful for children and adults.
> Planting trees is useless if they do not branch out wide enough.
> So, all the parks must be planted with trees.

6	WEAKEN

The intersection of Preston and Park roads..

Which one of the following undermines the argument the most?
A) School timing is the reason for accidents in other intersections as well.
B) There have been many dangerous accidents in this intersection around 8 AM even before any school was built in this area.

ANSWER

Answer: B

    The intersection of Preston and Park is very dangerous.
    Several accidents occur around 8 AM at this intersection.
    Several schools open during this time.
    Therefore, the school timing is the reason for the accidents.

A - incorrect - if school timing is the reason for accidents in other intersections as well, then it will only strengthen this argument.

B - correct - there have been many dangerous accidents around 8 AM at this intersection even before any school was built in the area.

This means that the schools in the area cannot be responsible for the accidents. So, this choice weakens the conclusion that school timing is the reason for the spate of accidents.

Answers
© Gift Of Logic, Inc * Copying prohibited

| 7 | WEAKEN |

Daniel lives in a neighborhood that is not..

Which one of the following undermines the argument the most?
A) The company that makes the burglar alarm that Daniel uses has warned recently that it is not one hundred percent reliable.
B) The burglar broke into Daniel's home through a rear window.

ANSWER

Answer: B

Add the choices to the argument and analyze.

A - incorrect – if the alarm is not one hundred percent reliable, then this fact strengthens Daniel's conclusion that it is defective.

B - correct - if the burglary was committed by entering through a rear window, that means the burglar alarm could not have sounded. Daniel only secured the rear door, not the rear window. So, this choice weakens Daniel's argument that the alarm was defective.

> Daniel lives in a neighborhood where there is crime.
> He secured the front and back doors with a burglar alarm.
> Opening the doors when alarm is set will cause the alarm to sound.
> Alarm did not sound yesterday when his home was burgled.
> The burglar broke into Daniel's home through a rear window.
> Therefore, Daniel concluded that there is a defect in the alarm.

Answers

8   WEAKEN

All the friends of Jasmine read one book in addition..

Which one of the following undermines the argument the most?
A) Jasmine always gets the best grade in each subject.
B) Jasmine can also get better grades if she reads one extra book in each subject.

ANSWER

Answer: A

Add the choices to the argument and analyze.

A - correct

   All the friends of Jasmine read one extra book for each subject.
   This has fetched them better grades.
   Jasmine always gets the best grade in each subject.
   Therefore, Jasmine must start reading one extra book for each subject.

If Jasmine always gets the best grade in each subject, then this fact weakens the conclusion that she must read one extra book for each subject. Her friends read the extra book for the purpose of getting better grades. Since Jasmine always gets the best grades, she does not have to read the extra books.

B - incorrect - if Jasmine can also get better grades by reading one extra book in each subject, then this fact supports the argument that she must start reading one extra book in each subject.

9	WEAKEN

Bank Manager: Recently, several clerks left their job..

Which one of the following undermines the Bank Manager's argument the most?
A) There was a long queue in the bank even before the clerks left their jobs.
B) It takes a lot of work to service the needs of bank customers.

ANSWER

Answer: A

A - correct - this choice weakens the argument.

> Recently, several clerks left their jobs.
> So, the bank does not have sufficient number of employees.
> There was a long queue in the bank even before the clerks left their jobs.
> So, there is a long queue in the bank these days.

If there was a long queue even before the clerks left their jobs (that is when the clerks were employed), then this fact undermines the Bank Manager's claim that the long queue is because of the clerks leaving their jobs.

B - incorrect - if it takes a lot of work to service the needs of bank customers, then this fact along with the fact that there are insufficient number of employees in the bank supports the conclusion that there is a long queue in the bank these days. So, this choice strengthens the argument.

Answers
© Gift Of Logic, Inc * Copying prohibited

10 WEAKEN

Astronomer: When I was star gazing last night..

Which one of the following undermines the Astronomer's argument the most?

A) Some shooting stars have the energy to reach the surface of Earth.
B) Shooting stars disintegrate before they reach the surface of Earth.

ANSWER

Answer: B

A - incorrect

Since this choice affirms that some shooting stars have the energy to reach the surface of Earth, it strengthens the conclusion that some stars can be expected to fall on Earth.

B - correct - this choice weakens the conclusion.
    I spotted twenty shooting stars last night.
    Shooting stars disintegrate before they reach the surface of Earth.
    Therefore, we can expect some stars to fall on Earth.

If shooting stars disintegrate before they reach the surface of Earth, they cannot fall on Earth as claimed. So, the conclusion is weakened.

11	WEAKEN

Restaurant manager: The lettuce in our salad..

Which one of the following undermines the argument the most?
A) Unwashed lettuce is the common reason for contamination of salads.
B) Bad tomatoes can contaminate the salad even if the lettuce is washed thoroughly.

ANSWER

Answer: B

A - incorrect - this choice indicates that unwashed lettuce is the common reason for salad contamination. But, the fact is that this restaurant uses only washed lettuce in their salads. So, this choice does not weaken the argument that there is no chance of contamination by bacteria.

B - correct- this choice weakens the argument.

> Lettuce in our salad is the best anyone can get.
> Farmer washes it before supplying it to us.
> We wash it again before using it.
> Bad tomatoes can contaminate the salad even if the lettuce is washed thoroughly.
> Therefore, there is no chance of contamination of our salad.

If there are bad tomatoes in the salad, it can contaminate the salad even if the salad is washed thoroughly. Therefore, this fact weakens the conclusion that there is no chance of contamination of the salad.

12   WEAKEN

Barber: Your hair is very curly. Most people..

Which one of the following undermines the argument the most?
A) A more expensive comb must be used to untangle curly hair.
B) Hair clipping machines can cut curly hair as quickly as they can cut straight hair.

ANSWER

Answer: B

A - incorrect - using a more expensive comb to untangle dense and curly hair will justify the barber charging extra money. So, this does not weaken the argument.

B - correct - this choice weakens the argument.

  Your hair is very dense and curly.
  Most people do not have this much dense and curly hair.
  Hair clipping machines can cut curly hair as quickly as they can cut straight hair.
  Therefore, I have to charge you extra money for cutting your hair.

If it takes hair clipping machines the same time to cut curly hair as it takes to cut straight hair, then the barber is not justified in charging extra money for the haircut.

13 WEAKEN

Car mechanic: Your car's engine must be replaced..

Which one of the following undermines the argument the most?
A) Abnormal amounts of fuel and air will damage the engine.
B) The oxygen sensor alone can be replaced with a new one.
C) When one part of a machine malfunctions, the entire machine must be replaced.

ANSWER

Answer: B. Note that the conclusion of this argument is in the first statement itself (the car's engine must be replaced).

A - incorrect - if abnormal amounts of fuel and air will damage the engine, this gives more strength to the argument that the engine must be replaced.

B - correct - this choice weakens the argument.
> The oxygen sensor in the engine is not working.
> This causes an incorrect amount of air and fuel to flow into the engine.
> The oxygen sensor alone can be replaced with a new one.
> Therefore, your car's engine must be replaced.

If the oxygen sensor alone can be replaced, then it weakens the conclusion of the mechanic that the entire engine must be replaced.

C - incorrect - if an entire machine must be replaced even if one part of the machine malfunctions, then this condition strengthens the mechanic's argument that the entire engine must be replaced because one part in it, namely, the oxygen sensor is not working.

14	WEAKEN

When flight# 30 left runway 22L at 9:00 AM ...

Which one of the following undermines the argument the most?
A) The vortex of air created behind an aircraft can destabilize another aircraft that comes too close to it.
B) The pilot of flight# 16 reported a problem in one of its engines.

ANSWER

Answer: B

A - incorrect - this strengthens the conclusion that Flight #16's problems were caused by flight #30.

B - correct
    Flight 30 - take off at 9 AM - did not report any problems.
    Flight 16 - take off at 9:02 AM - reported problems.
    The pilot of flight# 16 reported a problem in one of its engines.
    Therefore, flight #30 is responsible for flight #16's problem.

If the pilot of the flight #16 reported an engine problem, then this weakens the conclusion that flight # 30 is responsible for the problems of flight #16.

Answers

© Gift Of Logic, Inc * Copying prohibited

| 15 | WEAKEN |

Polluted water was proven to be the cause..

Which one of the following undermines the argument the most?
A) Water to the entire farm is supplied by only one lake.
B) People who ate cabbage grown in the farm did not have any health related problem.

ANSWER

Answer: B
Only the spinach grown in the farm have E-Coli bacteria. But, the conclusion is about avoiding purchase of all produce from the farm. Produce refers to all the things grown in the farm, not just the spinach.

A - incorrect - this strengthens the conclusion - if only one lake supplies water to the entire farm, and since the polluted water caused the E-Coli in spinach, we can conclude that all the produce grown in the farm will be contaminated. So, this choice strengthens the argument that all the produce from the farm must be avoided.

B - correct - this weakens the conclusion.

> Polluted water caused E-Coli in spinach.
> People who ate this spinach had kidney problems.
> People who ate cabbage grown in the farm did not have health problems.
> Therefore, the produce from the farm must be avoided.

Since people who ate cabbage did not have health problems, this choice weakens the argument that all the produce from the farm must be avoided.

Answers

16 WEAKEN

Hypertension is caused by high blood..

Which one of the following undermines the argument the most?
A) The sodium in the salt plays an important role in causing high blood pressure.
B) Some low-salt diets are not effective in reducing hypertension.

ANSWER

Answer: B
High blood pressure c→ hypertension
Excessive salt in kidneys c→ high blood volume
High blood volume c→ high blood pressure
Therefore, only diets low in salt are suitable for reducing hypertension.

The causal chain is:
excess salt c→ high blood vol c→ high blood pressure c→ Hypertension

A - incorrect - we can conclude from the given premises that excess salt in kidneys is likely to cause high blood pressure. This choice gives additional detail by indicating that the sodium in the salt plays and important role in causing high blood pressure. This does not weaken the conclusion in any way.

B - correct - this choice casts a doubt on the entire causal chain by claiming that certain low-salt diets do not reduce hypertension. So, the conclusion that only low-salt diets are suitable for reducing hypertension is questioned and weakened.

1 SUDOKU

Solve the following Sudoku. A correctly solved Sudoku has numbers 1-9 appearing only once in each row, each column and each 3x3 grid. Solving Sudokus will help you to gain valuable analytic skills.

5	8	7	4	1	2	3	9	6
6	4	1	5	3	9	8	7	2
2	3	9	8	7	6	5	1	4
8	7	5	2	9	3	6	4	1
1	2	3	6	4	8	7	5	9
9	6	4	7	5	1	2	8	3
4	1	2	3	8	7	9	6	5
3	5	8	9	6	4	1	2	7
7	9	6	1	2	5	4	3	8

Answers

© Gift Of Logic, Inc * Copying prohibited

2   SUDOKU

Solve the following Sudoku. A correctly solved Sudoku has numbers 1-9 appearing only once in each row, each column and each 3x3 grid. Solving Sudokus will help you to gain valuable analytic skills.

3	9	7	8	4	5	1	6	2
2	8	5	6	1	9	7	4	3
4	1	6	7	3	2	8	5	9
6	3	4	9	8	1	2	7	5
1	5	9	2	7	4	6	3	8
7	2	8	5	6	3	4	9	1
8	4	3	1	5	7	9	2	6
9	7	1	3	2	6	5	8	4
5	6	2	4	9	8	3	1	7

Answers

3

SUDOKU

Solve the following Sudoku. A correctly solved Sudoku has numbers 1-9 appearing only once in each row, each column and each 3x3 grid. Solving Sudokus will help you to gain valuable analytic skills.

4	6	3	5	2	1	7	8	9
7	9	2	4	3	8	6	5	1
5	8	1	9	7	6	4	2	3
1	3	5	8	6	7	9	4	2
8	4	6	3	9	2	1	7	5
9	2	7	1	5	4	8	3	6
6	1	8	2	4	5	3	9	7
2	7	9	6	8	3	5	1	4
3	5	4	7	1	9	2	6	8

Answers

© Gift Of Logic, Inc * Copying prohibited

4
SUDOKU

Solve the following Sudoku. A correctly solved Sudoku has numbers 1-9 appearing only once in each row, each column and each 3x3 grid. Solving Sudokus will help you to gain valuable analytic skills.

9	5	7	1	4	8	6	2	3
3	6	1	7	5	2	8	4	9
2	8	4	9	6	3	1	7	5
1	2	9	5	7	4	3	8	6
5	4	3	6	8	1	7	9	2
6	7	8	2	3	9	5	1	4
7	9	2	3	1	5	4	6	8
4	1	5	8	2	6	9	3	7
8	3	6	4	9	7	2	5	1

Answers

5

SUDOKU

Solve the following Sudoku. A correctly solved Sudoku has numbers 1-9 appearing only once in each row, each column and each 3x3 grid. Solving Sudokus will help you to gain valuable analytic skills.

5	2	1	9	3	6	7	8	4
8	7	6	4	5	1	9	2	3
3	9	4	8	2	7	5	6	1
4	6	7	2	1	9	3	5	8
1	8	9	5	6	3	2	4	7
2	3	5	7	4	8	6	1	9
9	1	3	6	8	2	4	7	5
6	5	8	3	7	4	1	9	2
7	4	2	1	9	5	8	3	6

Answers

1 POSITIONING

Diagram the scenario first and then answer the questions.

Tom, Dick, and Harry.. >> represent these as T, D, and H
Tom must sit immediately to the left of Harry >>TH

1) Can Tom sit in the third chair?
Answer: No. If Tom sits in the third chair, then Harry will not have a chair to sit on. See the second row of the diagram.

T,D,H
TH

	1	2	3	
			T	invalid
	T	H	D	valid
	H			invalid

2) Can Tom sit in the first chair?
Answer: Yes. If Tom sits on the first chair, then Harry will sit on the second chair (because of the TH rule) and Dick will sit on the third chair. See the third row of the diagram.

3) Can Harry sit in the first chair?
Answer: No. See the fourth row of the diagram. If Harry sits in the first chair, then Tom cannot sit to his left, since rule TH will be violated.

Answers
© Gift Of Logic, Inc * Copying prohibited

1	POSITIONING

4) Can Dick sit in the third chair?
 Answer: Yes. See diagram below. If Dick sits in the third chair, then Tom has to sit in the first chair and Harry in the second chair.

T,D,H
TH

	1	2	3
	T	H	D

5) Dick can sit in the first chair or the third chair.

Answer: True. If Dick sits in the first chair, then the second and third chairs can be taken by Tom and Harry. If Dick sits in the third chair, then the Tom and Harry will sit in the first and second chairs. Either scenario is possible.

T,D,H
TH

	1	2	3	
	D	T	H	valid
	T	H	D	valid

Answers

2 POSITIONING

Diagram the scenario.
Rolly, Polly, and Molly.. >> R,P,M

Note that the question sounds confusing because it uses names that are rhyming - like Rolly, Polly and Molly. Solving this problem will strengthen your concentration by focusing on the logic only.

Rolly and Polly must sleep next to each other >> RP ∦ PR. Note that either rule can be satisfied.

1) Can Molly sleep in bed# 2?
Answer: No. If Molly sleeps in bed# 2, then the diagram will look like the one shown below. In this setup, Rolly will not be able to sleep next to Polly. RP ∦ PR cannot be satisfied.

R,P,M
RP ∦ PR

1	2	3
R	M	P

2) Can Polly sleep in bed# 1?
Answer: Yes. If Polly sleeps in bed# 1, then Rolly will have to sleep in bed# 2 (to satisfy the PR rule) and Molly will sleep in bed# 3.

R,P,M
RP ∦ PR

1	2	3
P	R	M

Answers
© Gift Of Logic, Inc * Copying prohibited

2	POSITIONING

3) If Molly sleeps in bed# 1, where can Rolly sleep?
   A) bed# 2 only B) bed# 2 or bed# 3

Answer: B) bed# 2 or bed# 3 as shown below.

R,P,M
RP ∦ PR

1	2	3
M	R	P
M	P	R

3	POSITIONING

Vivek, Tom, Sandra, and Nazia are to be photographed >> V,T,S,N

Vivek must stand in the first spot >> V1

Sandra must stand in the third spot >> S3

1) Can a picture be taken in the following order?
   Vivek, Sandra, Tom, Nazia

V,T,S,N
V1
S3

1	2	3	4
V		S	

Answer: B) No. The diagram represents the position of the V and S. Sandra must be in the third spot, not in the second spot. So, V,S,T,N is not a valid positioning for taking pictures.

Answers

| 3 | POSITIONING |

2) Write the possible positions in which a picture can be taken.

	1	2	3	4
V,T,S,N				
V1	V	T	S	N
S3	V	N	S	T

With V and S fixed to positions 1 and 3, we have T and N available to take positions 2 and 4. So, the possible positions in which a picture can be taken are:

    Vivek, Tom, Sandra, Nazia
    Vivek, Nazia, Sandra, Tom

3) How many group pictures can be taken?

Answer: B)2.
This question can be answered by looking at the answer to the previous question.

Answers
© Gift Of Logic, Inc * Copying prohibited

4	POSITIONING

Three cars - a Toyota, a Honda, and a Ford .. >> T,H,F

The Toyota must be parked before the Honda >> T-H.

Note that this rule allows a car to be in between the Toyota and the Honda.

T,H,F
T-H

1	2	3
		T
F	T	H
T	H	F
H		

1) Can Toyota be in the third spot?
Answer: B) No. See the second row in the diagram. In this case, Toyoto cannot be in the third spot because there will be no way to park the Honda, which must be parked after the Toyota.

2) Can Toyota be in the second spot?
Answer: B) Yes, see the third row in the diagram.

3) Can Ford be in the first spot?
Answer: A) Yes, this can be verified from the third row in the diagram.

4) Can Toyota be in the first spot?
Answer: A) Yes. See the fourth row in the diagram. T1,H2,F3 is valid.

5) Can Honda be in the first spot?
Answer: B) No, See the fifth row in the diagram. If Honda is in the first spot, then Toyota cannot be parked before the Honda.

Answers

5 POSITIONING

A green, a black and a red shoe.. >> G,B,R

Black shoe must be placed after the red shoe >> R-B. Note that there can be a shoe between R and B.

G,B,R
R-B

1	2	3
R	B	G
R	G	B
G	R	B

1) If the green shoe is placed in the third position, then the black shoe must be placed in which position?

Answer: B) 2. See the second row in the diagram. If G is placed in the third spot, that leaves spots 1 and 2 for R and B. To place R and B according to rule R-B, R has to be in spot 1 and B in spot 2.

2) If the green shoe is placed in the second position, then the black shoe must be placed in which position?

Answer: B) 3. See the third row in the diagram.

3) If the green shoe is placed in the first position, then the black shoe must be placed in which position?

Answer: B) 3. See the fourth row in the diagram.

Answers
© Gift Of Logic, Inc * Copying prohibited

6 POSITIONING

Three boats are to sail in the Mississippi river..>> B1, B2, B3

B2 must sail immediately before B1 >> B2B1

B1,B2,B3
B2B1

1	2	3
B3	B2	B1
B2	B1	B3
		B2

1) If B2 sails in the middle, then B3 can sail in which position?
Answer: A) 1. See the second row in the diagram. If B2 sails in the middle, B1 has to sail at the end (to satisfy the B2B1 rule), leaving B3 to sail first.

2) If B2 sails first, then B3 can sail in which position?
Answer: B) 3. See the third row in the diagram. B1 has to sail in the middle because of rule B2B1. So, B3 must sail in the third position.

3) Can B2 sail in the last position?
Answer: B) No. See the fourth row in the diagram. This will violate the B2B1 rule which states that B2 must be immediately before B1 or B1 must be immediately after B2.

Answers
© Gift Of Logic, Inc * Copying prohibited

7 POSITIONING

Represent Abra, Babra, and Cabra with A, B and C respectively.

If Abra walks in the first position, then Babra must be in the third position.
>> A1 → B3
If Babra is in the second position, then Cabra must be in the first position.
>> B2 → C1

1) Can the donkeys walk in the following order?
   Abra, Babra, Cabra

Answer B) No. In the given order, Abra is in the first position. So, according to rule A1 → B3, Babra must be in the third position, not the second.

2) Can the donkeys walk in the following order?
   Cabra, Babra, Abra

Answer: A) Yes. Abra is in third position. There is no rule regarding Abra being in third position. Rule B2 → C1 is satisfied with Babra in second spot and Cabra in the first spot.

3) Can the donkeys walk in the following order?
   Cabra, Abra, Babra

Answer: A) Yes. Rule A1 → B3 does not apply because A is not in spot 1, but it is in spot 2. Rule B2 → C1 also does not apply because B is in position 3.

Answers
© Gift Of Logic, Inc * Copying prohibited

8 POSITIONING

Johnny wants to wear three watches ..

1) Verify whether each of the following choices could represent the possible positions of the watches that Johnny can wear from first to last.

W2 cannot be the first watch >> ~W2@1. The @ symbol is used to represent the word "at". Keeping this in mind, it is easy to go through each of the following options and say which one is correct.

A) W2,W1,W3 >> violates the ~W2@1 rule
B) W3,W1,W2 >> possible
C) W1,W3,W2 >> possible
D) W2,W3,W1 >> violates the ~W2@1 rule
E) W3,W2,W1 >> possible
F) W1,W2,W3 >> possible

2) If W3 also should not be worn first, then write the possible ways in which the watches can be worn.

In addition to the ~W2@1 rule, now we have a ~W3@1 rule. Since W2 and W3 can not be the first watch, W1 only can be the first watch. W2 and W3 can be either the second or the third watch.

    W1 W2 W3 and W1 W3 W2 are the two possibilities.

Answers
© Gift Of Logic, Inc * Copying prohibited

9 POSITIONING

Three books, labeled B1, B2, and B3 ..

B1 can be in position 1 or in position 2, but it cannot be in position 3.
    >>   B1@1⊦ B1@2, ~B1@3

B3 must not be in position 2 >> ~B3@2

B1,B2,B3
B1@1⊦ B1@2
~B1@3
~B3@2

1	2	3
B2	B1	B3
B1	B3	B2
B3	B2	B1

1) Is the following ordering of books correct? B2, B1, B3
Answer: A) Yes. See the second row in the diagram. None of the rules are violated.

2) Is the following ordering of books correct? B1, B3, B2
Answer: B) No. See the third row in the diagram. Rule ~ B3@2 is violated.

3) Can the books be placed in the following order?
    B3, B2, B1
Answer: B) No. See the fourth row in the diagram. The rule ~B1@3 is violated in this case.

Answers
© Gift Of Logic, Inc * Copying prohibited

10 POSITIONING

Abigail must wear a blue or green shirt ..

The rules can be represented as follows:
    B ∦ G everyday
    ~same color on consecutive days

1) Can Abigail wear shirts in the order shown for each week? Write Yes/No in each row in the column captioned as 'Possible'.

Week	Mon	Tue	Wed	Thurs	Fri	Possible?
1	Blue	Green	Blue	Blue	Green	No (same color on wed and thurs)
2	Blue	Green	Blue	Green	Blue	Yes
3	Green	Blue	Red	Green	Blue	No (only Blue or Green allowed)
4	Green	Blue	Green	Blue	Green	Yes

---

11 Amber must wear ...

The rules can be represented as follows:
  B ∦ G everyday
  same color on consecutive days ok

1) Can Amber wear the skirts in the order shown in the table below? Write Yes/No in each row in the column captioned as 'Possible'.

Week	Mon	Tue	Wed	Thurs	Fri	Possible?
1	Blue	Green	Blue	Green	Blue	Yes
2	Blue	Blue	Blue	Blue	Red	No (only Blue or Green allowed)
3	Green	Green	Blue	Blue	Green	Yes

answers

© Gift Of Logic, Inc * Copying prohibited

12 POSITIONING

Kate can wear a blue or a green skirt ..

The rule can be represented as follows:
B ∦ G - Mon
O ∦ Y - Tue
R ∦ G - Wed

1) Which of the following choices are possible? Write Yes/No in the column marked "Possible" for each of the choices shown.

Applying these rules for each row helps to answer the question. The rule that is violated in each choice is shown in parenthesis.

Monday	Tuesday	Wednesday	Possible?
Orange	Yellow	Green	No (B ∦ G -Mon)
Green	Orange	Green	Yes
Blue	Green	Orange	No (O ∦ Y -Tue)
Green	Yellow	Green	Yes

Answers
© Gift Of Logic, Inc * Copying prohibited

13 POSITIONING

Mike plays soccer and hockey ..

The rules can be represented as follows:
  S & H - Wed
  H || B - Thurs
  Sk ⫫ Sw - Friday

Note that he can play Hockey or Baseball on Thursday. This means, he can play Hockey only or Baseball only or both Hockey and Baseball. The rule H || B does not state that he can not play both. The rule Sk ⫫ Sw - Friday states clearly that he cannot do both on Friday.

1) Which of the following scenarios are possible?

Applying the rules to each day helps in answering whether the choices are possible or not. The violations are shown in parenthesis.

Wednesday	Thursday	Friday	Possible?
Hockey and Baseball	Soccer	Skating	No (S & H - Wed)
Soccer and Hockey	Baseball	Skating	Yes
Skating and Swimming	Hockey	Soccer	No (S & H - Wed)
Soccer and Hockey	Hockey and Baseball	Skating and Swimming	No (Sk ⫫ Sw)
Hockey and Soccer	Baseball	Swimming	Yes

Answers

GROUPING

1

A box contains the following..

Write below all the possible ways in which you could pick any two objects from the box.

    paper, pencil
    pencil, eraser
    eraser, paper

---

2 Write all the possible combinations of animals that Billy could choose as pets.

    cat, dog
    dog, mouse
    mouse, cat

---

3 Represent Sofa as S and Chair as C. The rule "Sofa and chair must not be purchased together" can be represented as ~SC. Apply the rule to each of the choices below. Note that since the order does not matter, ~SC means ~CS as well.

1) Which of the following purchases can be made?
A) sofa, table >> Correct
B) table, chair >> Correct
C) chair, sofa >> Incorrect, violates the ~SC rule.
D) chair, mirror >> Correct

Answers
© Gift Of Logic, Inc * Copying prohibited

GROUPING

4   Toyota, Honda, Mazda >> T,H,M

Select two cars from the above list.
Toyota and Mazda must be selected together. >> TM

1) Which of the following selections can be made?

A) Toyota, Honda - incorrect, violates the TM rule.

B) Honda, Mazda- incorrect, violates the TM rule.

C) Mazda, Toyota- correct, satisfies the TM rule.

---

5   Carmen bought a pack of six balloons...

Remember the rule "The two red balloons must be in the same bunch" and apply this rule to each answer choice .

1) Which of the following choices have valid bunches?

A >> Correct. The two red balloons are in the same bunch.

B >> Incorrect. Even though the two red balloons are in the same bunch, there are three blue balloons in this choice whereas there are only two blue balloons totally.

C >> Correct. The two red balloons are in the same bunch.

Answers
© Gift Of Logic, Inc * Copying prohibited

GROUPING

6

From a pack of six balloons...

Remember the rule "A green balloon must be in each bunch" and apply this rule to each answer choice. Also, make sure that the number of balloons does not exceed their limits (1-red, 2 blue and 3-green).

1) Which of the following choices show the correct color of the balloons?

A) Green, Red   Blue, Blue   Green, Green
   >>Incorrect, green balloon is not in bunch-2.

B) Green, Red   Green, Blue   Green, Blue
   >> Correct.

C) Blue, Blue   Green, Red   Green, Green
   >> Incorrect, green balloon is not in bunch-1.

Answers
© Gift Of Logic, Inc * Copying prohibited

GROUPING

7 Four balls on the table...

1) Pick one red ball and one green ball from the table. Write all the possible ways in which the balls can be selected.

    Red1, Green3
    Red1, Green4
    Red2, Green3
    Red2, Green4

2) Pick two red balls and one green ball from the table. Write all the possible ways in which the balls can be selected.

    Red1, Red2, Green3
    Red1, Red2, Green4

3) Pick two green balls and one red ball. Write all the possible ways in which the balls can be selected.

    Green3, Green4, Red1
    Green3, Green4, Red2

Answers

GROUPING

8 Four flowers...
Diagram the rules as follows.
The red and the blue flowers must not be selected together. >> ~RB
The green and yellow flowers must not be selected together. >> ~GY

1) Which of the following is a correct selection?
    A) Red, Blue >> incorrect - violates the ~RB condition
    B) Blue, Red >> incorrect - violates the ~RB condition
    C) Red, Yellow >> correct answer
    D) Yellow, Green >> incorrect - violates the ~GY condition

---

9 Four flowers... Select three flowers..

Diagram the conditions as follows.
The red and blue flowers must not be selected together. >> ~RB
The red and yellow flowers must not be selected together. >> ~RY

1) Which of the following is a correct pick?
    A) Red, Green, Yellow >> incorrect - violates the ~RY condition.
    B) Red, Green, Blue >> incorrect - violates the ~RB condition.
    C) Red, Yellow, Blue >> incorrect - violates the ~RB condition.
    D) Green, Blue, Yellow >> correct answer

Answers
© Gift Of Logic, Inc * Copying prohibited

GROUPING

10 Following are the contents...

1) Which of the following is a correct selection?
A) apple1, apple2, orange3, apple3 >> incorrect - there should be two oranges.
B) apple1, apple3, orange1, orange2 >> correct answer

---

11 Following are the contents.. Fruits of the same kind...
Which of the following is a correct selection?
A) apple1, apple2, orange3, orange1 >> incorrect- apple1 and apple2 are from same box.
B) apple1, apple3, orange3, orange4 >> incorrect - orange3 and orange4 are from the same box.
C) apple1, apple2, orange1, orange2 >> incorrect - apple1 and apple2 are from the same box.
D) apple1, apple4, orange2, orange3 >> correct answer - apples are from different boxes and oranges are from different boxes.

---

12 A sports utility box..
Pick two bats and three balls and write your selection below.

baseball bat; ping-pong bat; ping-pong ball, tennis ball; soccer ball

Note that there are four balls, but you must pick only three. The above selection leaves out the golf ball. We can make an alternate selection that includes the golf ball, but leaves out some other ball.

Answers
© Gift Of Logic, Inc * Copying prohibited

GROUPING

13
parrot, lion, cat, butterfly, tiger, horse, eagle, dove

From the list above, select two birds and at least two animals. Write your selection below.
  parrot, butterfly, lion, cat, (tiger), (horse)
The animals shown in parenthesis are optional selections.

14
car1, truck1, car2, truck2, car3, truck3, truck4

Select no more than two cars and no less than three trucks from the list shown above.
1) What is the least number of cars that can be picked? Answer: 0
2) What is the most number of cars that can be picked? Answer: 2
3) What is the least number of trucks that can be picked? Answer: 3
4) What is the most number of trucks that can be picked? Answer: 4

---

15   triangle1, circle1, triangle2, circle2, triangle3, circle3

Select at most two circles and at most three triangles from the list shown above.

1) What is the least number of circles that can be picked? Answer: 0
2) What is the most number of circles that can be picked? Answer: 2
3) What is the least number of triangles that can be picked? Answer: 0
4) What is the most number of triangles that can be picked? Answer: 3

Answers

© Gift Of Logic, Inc * Copying prohibited

PATTERN PERCEPTION

Question#	Answer
1	B
2	A
3	B
4	A
5	B
6	A
7	B
8	A

FIGURE FORMATION

Question#	Answer
1	B
2	A
3	B
4	B
5	A
6	B
7	A
8	A

Answers
© Gift Of Logic, Inc * Copying prohibited

PAPER FOLDING AND CUTTING

Question#	Answer
1	A
2	A
3	B

FIGURE MATRIX

Q#	Answer	Reasoning
1	B	boy will become a man; girl will become a woman
2	A	cow gives milk; chicken gives eggs
3	A	nail is driven with a hammer; screw is driven with a screwdriver
4	B	airplane has a pilot; space shuttle has an astronaut
5	A	cow, and the other three animals are herbivorous
6	C	ring is an ornament, like the three other ornaments
7	C	stove is used for cooking, similar to the others
8	C	harmonica is played with the mouth, like the others

RULE DETECTION

Question#	Answer	Question	Answer
1	A	4	C
2	B	5	A
3	A	6	B

Answers
© Gift Of Logic, Inc * Copying prohibited

NOTES

www.ingramcontent.com/pod-product-compliance
Lightning Source LLC
Chambersburg PA
CBHW080301180526
45167CB00006B/2623